IMAGES
of Aviation

US NAVAL AIR STATION
GROSSE ILE

The patch for Naval Air Station Grosse Ile featured a winged textbook. A star showed its location in Michigan. The patch, created by a base artist, was selected as the winning entry in a contest. (Courtesy of the Grosse Ile Historical Society.)

ON THE COVER: This is a 1949 photograph showing row after row of Navy Reserve FG-1D Corsairs parked on the ramp below the base's control tower. (Courtesy of the Grosse Ile Historical Society.)

IMAGES
of Aviation

US NAVAL AIR STATION
GROSSE ILE

Kenneth M. Keisel and
the Grosse Ile Historical Society

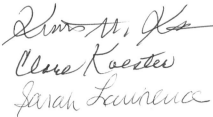

ARCADIA
PUBLISHING

Published by Arcadia Publishing
Charleston, South Carolina

Printed in the United States of America

Library of Congress Control Number: 2011933882

For all general information, please contact Arcadia Publishing:
Telephone 843-853-2070
Fax 843-853-0044
E-mail sales@arcadiapublishing.com
For customer service and orders:
Toll-Free 1-888-313-2665

Visit us on the Internet at www.arcadiapublishing.com

This book is dedicated to every man and woman who served in any capacity at Naval Air Station Grosse Ile—their legacy is not forgotten. If this book serves to rekindle their memories and names, it will have succeeded. It is also dedicated to the residents of Grosse Ile and the downriver Detroit area, who for 40 years lived with an active naval air station in their own backyard. This is your story, too, and there is much here to be proud of. It is also dedicated to the staff and volunteers of the Grosse Ile Historical Society and the Grosse Ile Kiwanis, who maintain the Base Museum and Memorial Garden.

CONTENTS

Acknowledgments   6

Introduction   7

1.   The Early Years   11

2.   The ZMC-2   29

3.   The War Years   37

4.   The Postwar Years   85

5.   Accidents   107

6.   Grosse Ile Today   117

ACKNOWLEDGMENTS

No book on a subject as large as Naval Air Station Grosse Ile (NASGI) comes about without the dedication and hard work of a large number of people, and every one of you deserves to be recognized for your tireless efforts. You are custodians of the legacy of NASGI, and your contributions are the only reason this book was possible. A big thank you goes out to Clare Koester of the Grosse Ile Historical Society (GIHS), who assembled this book while I typed away—you are an angel. Your constitution and endurance are the stuff of Olympic legend. To Sarah Lawrence at the GIHS for digging through hundreds of rough photographs looking for the gems. To Anna Wilson at Arcadia Publishing for jumping at the chance to tell NASGI's story. I hope we have produced something worthy of your enthusiasm. To Melvin Smith, who served at the D-51 Nike Missile Site, you made my explanation not only accurate but also possible. To John Ansteth, who told me about his grandfather's adventures as the first pilot of the ZMC-2. To Hayden Hamilton and the staff of the American Aviation Historical Society (AAHS), who taught me everything I know about the McDonnell FH-1 Phantom. To director Derek M. Thiel and the staff of the Grosse Ile Township Airport, thank you for the guided tour. Hope you finally got all the mud out of your car. To the staff of the National Archives, who searched through millions of photographs to find "that one with the *Hindenburg*." Thanks to Stanley Outlaw and Hal Neubauer, who supervise the NASGI online community. Posthumous thanks go to Lt. Comdr. Dick Melton, USN, whose book *The Forty Year Hitch* never let me down, and to Walker C. Morrow and Carl B. Fritsche, inventors of the ZMC-2 and authors of the invaluable book *The Metalclad Airship ZMC-2*. Posthumous thanks also go to Don Fuller, base veteran and long-time keeper of everything NASGI. Don, you truly were "the wind beneath our wings." Finally, thanks to my parents, especially to my father Kenneth G. Keisel USAF ret. Dad, I bet you never thought you'd see your name in a book about the Navy!

All images, except where otherwise noted, are from the collection of the Grosse Ile Historical Society.

INTRODUCTION

Naval Air Station Grosse Ile was one of the most important, and most unusual, military bases ever constructed on American soil. Known as "Detroit's unsinkable aircraft carrier" during the height of World War II, it was the Navy's largest primary flight training base. This raises the question, who served at NASGI, and why did they go there? Suppose, for a moment, that it is 1942, you have just joined the Navy wanting to fly. Unlike most fellows who join the Navy, you do not want to serve on a destroyer or a submarine—you want wings. Having the option of joining the Army Air Corps (there was no Air Force until 1947), you instead joined the Navy, and no one in the Navy is ever going to question that decision. You don't want to fly from an air base but want to take off and land from an aircraft carrier—a floating runway that from the air looks about the size of a postage stamp sitting in the middle of the ocean. Worse yet, it is moving, and not just forward, but up and down as well. You have decided to become what is generally considered the toughest job in the military—to be a Navy carrier pilot. And Grosse Ile is where carrier pilots are made.

You already know a few things. First, you won't be shooting down any German planes; the Army pilots are taking care of that problem. You are going to be learning to bank an airplane into a steep dive to drop bombs onto the decks of enemy warships; maybe the same aircraft carriers that launched the planes at Pearl Harbor. If you are very fortunate, you might get to fly the fighters that protect the planes dropping bombs. Second, the chances of getting hurt or killed in training are far greater than you would have faced having gone to Norfolk or San Diego to serve on destroyers and submarines. You are not going be trained in knot-tying or giving orders to seamen if you have made it to Grosse Ile. There aren't any training launches or periscopes here. There are rudders and propellers, but they are not attached to ships. Your training is of the highest level—the most precise, most careful, and most dangerous of its kind. By the time you have completed training, you will have learned that going left instead of right is not going to mean "trading paint" with a boat next to you, it will result in colliding with the fellow flying formation alongside, which will almost certainly be fatal. You have come to Grosse Ile so the Navy can see if you have what it takes to become a carrier pilot, to see if you have what author Tom Wolfe calls "the right stuff."

If you are very fortunate, you are arriving at Grosse Ile by air in a transport plane. Most likely it as an R4D, the Navy's version of the C-47 Gooney Bird. It is probably the first airplane you have ever been on, possibly the first you have ever been close to. If you are smart enough to look out the window masking the final approach (and most pilots do), you will see your destination. It is an island, and from the air it is difficult to make out the size, but its shape and position in the river bears a strong resemblance to Manhattan Island in New York City. It is long and narrow, and it dominates the entrance to the river it splits in two, but instead of having the East River to its east and the Hudson River to the west, Grosse Ile has the Detroit River and the Trenton Channel, respectively. Like Manhattan, Grosse Ile is surrounded by several smaller islands. There is even a similar road running along the eastern shoreline, though instead of bearing the name East River Drive, it is called East River Road. What you see next will depend on what time of year it is. If it is winter, which in Michigan runs from November through March, everything seen will be white with snow. There are huge pieces of ice floating down the river past the islands looking like giant puzzle pieces waiting to be reassembled.

You are most likely approaching Grosse Ile from the south, flying over Lake Erie. To the left, just a few miles south of Grosse Ile, is the city of Toledo, Ohio. Down there, the Willys-Overland factory is working overtime producing thousands of Jeeps, the rugged little transports that will soon be carrying Americans through the hedgerows of Normandy and France on their way to defeat Germany. Seventy years later, they will still be manufacturing Jeeps in Toledo. A little further north and west is a large airport. That is Willow Run, where the Ford Motor Company is producing one B-24 Liberator bomber every hour. North of that, at a bend in the river, is the city of Detroit, where the same River Rouge Plant that gave birth to the Ford Model T is now employing 100,000 workers in production of war material. In Warren, the Detroit Tank Arsenal is building some of the 22,500 tanks it will assemble during the war. You are landing in the very heart of what President Franklin D. Roosevelt referred to as "the great arsenal of democracy." Down there, the Great Depression is over, spirits are high, and unemployment is virtually zero. To the right, just across the mile-wide Detroit River, is a wooded coastline. That is Ontario, Canada, and it is doing its part as well. On June 6, 1944, hundreds of Canadians will lose their lives in Normandy landing on Juno Beach, the second-most heavily defended landing after Omaha. Thousands more will be killed in the liberation of northern Belgium and Holland. The Dutch royal family has taken refuge in Ottawa, and Dutch princess Margaret will be born there in a hospital room declared "extraterritorial." Even today, it is hard for a Canadian veteran to pay his own bar tab in Amsterdam or The Hague. In the coming weeks, you may even find yourself paying an "unofficial" visit across the river to the popular bars and clubs of Amherstburg.

About to make a landing, you now see that Grosse Ile is no Manhattan. It is barely eight miles long and a mile and a half at its widest. Naval Air Station Grosse Ile dominates nearly a fifth of the island. The air station itself is instantly recognizable by its unique triangle of runways, pointing like an arrow south to Lake Erie. Near the middle of the triangle is a huge blimp hangar where the Navy's ZMC-2 was built. The ZMC-2 was, and still is, the world's only all-metal airship. As you make the landing, the reason you are here is clear. There is a row of bright yellow Boeing Stearman biplanes waiting on the flight line, but you are still weeks away from getting into one. There are still a lot of classes and reading before you will be entrusted to even sit in an airplane, but you have arrived at Naval Air Station Grosse Ile. It is the first step in proving to the world that you have what it takes to be a Navy carrier pilot—that you have the right stuff.

This story cannot be finished without adding a little about the island itself. During training there, what will you, as a young airman, discover about Grosse Ile? Once off the base, you will find that you are in a most unusual location for a military station. Grosse Ile is the little island with the big houses. As early as the mid-19th century, prosperous Detroit families had begun building lavish estates on Grosse Ile. Like most islands, it was developed from the outside in, with the first mansions appearing along the shoreline facing Canada or Trenton. By the early 20th century, Grosse Ile was home to many prominent families. The land that Naval Air Station Grosse Ile resides on was purchased from Ransom E. Olds, the founder of Oldsmobile. Indeed, the bay where the Navy built its first hangar is known as Olds Bay, and the mansion that served as the USO Club during the war was Olds's summer residence. How odd it must have been when the base was active to be sitting on the porch of an elegant Victorian mansion, where General Custer once visited (he grew up in nearby Monroe, Michigan), only to have this idyllic setting disturbed hourly by the roar of powerful Navy aircraft flying down the river just above the waves. That is not to say the Navy was not welcome on Grosse Ile—quite the contrary. During World War II, many islanders rented out spare rooms to Navy personnel when space on the base was in short supply. They interacted daily with thousands of military men occupying their island, but there is little evidence this was ever a nuisance. Until the arrival of Phantom jets in 1950, there were no complaints about the aircraft. When an occasional pilot parachuted into the trees at the local church, all that was asked for was his escape handle for their display case. The people of Grosse Ile are to be congratulated for their years of dedication to military service just as much as the base's personnel. In the end, it really is their story too.

For 40 years, Grosse Ile was the little island with the big houses—and one very big base.

Pictured here is a base parking sticker from the 1960s.

One

THE EARLY YEARS

The story of Naval Air Station Grosse Ile begins in July 1925 in Detroit, when a Naval Reserve lieutenant named Robert Bridges formed a small unit of four men who shared a common interest in the new field of naval aviation. They began meeting in the old Naval Armory a few miles east of downtown Detroit. They had no aircraft, so the group had to be content by studying instructions on flight theory and aircraft operations from books and Navy manuals. The passage of the National Defense Act in 1925 expanded Navy Reserve Aviation, and the group became official near the end of 1925. In 1926, the unit was designated as Torpedo Squadron VT-31 despite having no aircraft. Lt. Charles David Williams was assigned as its first commanding officer. In March 1927, VT-31 finally acquired its first airplane, a Consolidated NY-1 trainer, and moved out of the Naval Armory to Selfridge Field, an Army airbase near Mount Clemens, Michigan.

In 1927, VT-31 moved into a newly constructed hangar known as the "Tin Hangar" in Memorial Park in downtown Detroit. The NY-1 was brought down, and a single float was added. Soon the squadron's pilots were flying from the Detroit River near Belle Isle. All this activity in Memorial Park aroused concern from Detroit residents. A new permanent home for VT-31 was needed.

The solution lay at Grosse Ile. In late 1927, Lieutenant Williams was able to convince the Michigan Legislature to lease a five-acre area of cattail marsh on the southern tip of Grosse Ile adjacent to Olds Bay. Work draining the marsh continued through 1928. Even the old Tin Hangar was disassembled and brought down the Detroit River on a barge to be re-erected on the new site. In July 1929, the base was ready for operations, and US Naval Reserve Aviation Base Grosse Ile was officially dedicated on September 7, 1929. When a nearby flight school closed in 1932, Lieutenant Williams was able to obtain its property for the Navy. He did the same again in 1933, leasing all 375 acres owned by the Aircraft Development Corporation. In August 1935, the same year the Reserve Base at Grosse Ile received its new job to train Navy cadets, Lieutenant Williams was ordered to a new command on another island—Pearl Harbor, Hawaii. On December 7, 1941, Lieutenant Williams witnessed firsthand the start of World War II, and thanks to him Grosse Ile was ready.

Perhaps the earliest aerial photograph of Grosse Ile, this picture was taken around 1925 from an altitude of two miles. It shows the southern tip of the island as it looked before the arrival of the Navy in 1927. The location that would become Naval Air Station Grosse Ile is still covered with woods, scrub, and cattail marshes, while the rest of southern Grosse Ile is farmland. In the early 1900s, Ransom Olds, founder of Oldsmobile, purchased virtually the entire southern tip of Grosse Ile as well as much of Elba Island from the Groh family, who had owned and farmed it throughout the late 19th century. Olds built the largest home on the islands on the eastern shore of Elba, which he used as a summer retreat. In 1925, he sold 403 acres of Grosse Ile, as well as his property on Elba, to the Aircraft Development Corporation. Planning to construct innovative all-metal blimps on the site, the corporation developed it into the Grosse Ile Airfield. In May 1925, the Olds Mansion on Elba was converted into a pioneer air club called Chateau Voyageur.

The father of naval aviation on Grosse Ile, Lt. Charles David Williams Jr. was an excellent choice to take command of the Navy's new Air Reserve squadron. An experienced pilot, by 1925 he had been actively promoting naval aviation in the Detroit area for over a year. He also had connections in the Michigan Legislature that would prove essential in acquiring property on Grosse Ile's Olds Bay for the Navy in 1927 and again in 1932, when additional property became available on Groh Road. Under Lieutenant Williams's guidance, Detroit's Naval Air Reserve Torpedo Squadron VT-31 evolved from a half dozen inexperienced pilots without an airplane to Grosse Ile Naval Air Reserve Station, a fully functional instillation ready to do its duty when World War II arrived. Lieutenant Williams left Grosse Ile in 1935 to take command of another Island base—Pearl Harbor, Hawaii. He was there when the Japanese attacked in 1941.

This is the aircraft that brought naval aviation to the Detroit area, the Consolidated NY-1. The photograph was probably taken around 1927 at Selfridge Army Airfield near Mount Clemens, Michigan. When the Navy designated its Air Reserve squadron in Detroit as Torpedo Squadron VT-31, the squadron still had no aircraft and had to content itself with traveling to Naval Air Station Glenview near Chicago to get flight time. After nearly a year without an aircraft, the Navy assigned this aging NY-1 to the squadron, which then moved from the Naval Armory in Detroit to Selfridge Army Airfield. The NY-1 was no fighter. Slow and boxy, it was primitive compared to the sleek Curtiss P-1 Hawk fighters, which were flown by the Army at Selfridge. Still, it did its job, giving Navy Reserve aviators their first airplane. In 1927, the squadron would move back to Detroit, possibly because of an unfortunate incident when a Navy Air Reserve pilot accidentally cut the base commander's Curtiss P-1 nearly in half with the NY-1's propeller.

Consolidated Training Plane

The NY-1, now a floatplane, is lowered into the Detroit River from the Navy's temporary home in Memorial Park, near Belle Isle. Torpedo Squadron VT-31 moved from Selfridge Airfield to Memorial Park in summer 1927, building a small structure, called the Tin Hangar, in Memorial Park. The VT-31 remained in Downtown Detroit for only two years, returning to Selfridge Airfield during the winters.

In July 1929, the Navy base at Grosse Ile was ready. A year later, the reservists traded their aging NY-1s for a Keystone/Loening OL-9 floatplane and four Curtiss O2C-1s, a trainer variation of the Curtiss F8C-4 Helldiver that was often assigned to Reserve air stations. Other aircraft based at Grosse Ile during this era were Curtiss TS-1s, Martin T4Ms, and Boeing F4B-4s.

This photograph, taken around 1927, is among the earliest images of the original Grosse Ile Airfield. Looking east, it shows the 3,000-foot-wide circle cut by the Aircraft Development Corporation for the construction of its all-metal blimp, the ZMC-2. The blimp's hangar is on the far right, and the hangar built by Wings Inc. as a flight school is visible on the left. Groh Road is seen along the left edge. In 1928, the circle would be paved and named Meridian Circle Road. Connected to Meridian Road at Groh Road, the circle would be a popular destination for local residents who could drive cars and bikes around the airfield. Note the few houses existing at the time on East River Road and the northern half of Elba Island. (Courtesy National Archives.)

This c. 1930 photograph shows Grosse Ile Airfield looking southeast. The flight school, now owned by Curtiss-Wright, is seen at the bottom adjacent to Groh Road. The blimp hangar is at top right surrounded by trees. Naval Reserve Air Station Grosse Ile is the small cluster of buildings on Olds Bay just to the left of the blimp hangar. Elba Island is at top, with the Olds Mansion dominating its eastern shoreline.

This c. 1930 photograph shows Grosse Ile Airfield looking northeast. The area labeled as "A" is the Curtiss-Wright Flight School, area "B" is Meridian Circle Road, area "C" is the Naval Air Reserve Base, area "D" a rifle range, area "E" is the ZMC-2 blimp hangar, and area "F" is the Navy's original Tin Hangar. After being floated down the Detroit River, the Tin Hangar was reassembled on this site, where it would remain until moved near Groh Road in 1942, making way for the addition of runways.

This is the earliest known photograph of Naval Reserve Air Base Grosse Ile and was probably taken in late 1929. Under the advice of Lieutenant Williams and Lt. Comdr. Richard Thorton Broadhead, commanding officer of the Detroit Naval Armory, the State of Michigan purchased a five-acre finger of land from the Detroit Aircraft Corporation in the latter part of 1927, using it for the Michigan Naval Force. The property was mostly submerged cattail marsh on which Lt. Col. Broadhead secured $100,000 for construction of a seaplane base. In addition to the seaplane hangar, a repair shop, living quarters, a mess hall, and other buildings were constructed. Because of restrictions of the toll bridge, most of the building material had to be floated down the river on barges. Construction began in early 1928; it was ready in July 1929.

In 1932, the Curtiss-Wright Flight School fell victim to the Great Depression. The Michigan Legislature was not in session, so Lieutenant Williams borrowed funds from a fellow officer to pay back taxes on the facility, securing it for the Navy until the legislation resumed. After moving to the 11-acre flight school, this area became known as the "lower base." Navy aircraft were hangared in the former flight school on Groh Road and serviced in the seaplane hangar. The two sites were not connected, so Navy aircraft had to use the eastern half of Meridian Circle, a public road, to taxi between the two sites. In 1933, the State of Michigan began leasing all 375 acres of the Aircraft Development Corporation's property, finally giving the Navy complete control over the site. (Courtesy National Archives.)

Here are the Curtiss-Wright Flight School and adjacent barracks. The barracks were soon filled with Navy reservists as well as the base's station keepers. In the late 1920s, there were 17 station keepers assigned to Grosse Ile. By 1935, that number had doubled. Meridian Road was completed in 1925, and it is visible cutting through the woods at the top right.

This c. 1943 image shows the same building seen in the top photograph. The flight school barracks have become the base's Officers' Club and Bachelor Officer Quarters (BOQ). The base name has been transferred from the seaplane hangar to the roof of Hangar Two, and additional support buildings have been constructed. Four Boeing-Stearman Model 75 trainers are huddled nearby. Meridian Road, completed in 1925, is visible running north near the top. (Courtesy National Archives.)

This photograph shows winter on Grosse Ile as a Curtiss O2C-1 sits parked on the ramp at the lower base. Around 1930, Torpedo Squadron VT-31 changed its designation to VN-9. On March 11, 1930, Marine Corps Service Company SS-2MR was commissioned, and on June 19, 1930, Marine Squadron VO-5MR came into being. This picture was likely taken around 1931 before the Navy moved to Groh Road.

This photograph shows a lone Grumman FF-2 Fifi and O2C1s parked on Meridian Circle Road. Aircraft not stored in hangars were usually parked on the flight line in front of the flight school. In 1929, flush runway lights were installed. They were activated automatically by weather-vane control to outline the runway most directly into the wind—a first in the nation.

Cadets and officers pose for a photograph in front of a Grumman FF-2 Fifi fighter. The cadet's neckties betray strong winds from the west even though the propeller is perfectly positioned. The tree line along Frenchman's Creek is visible in the background.

A formation of Grumman FF-2 flies over the Detroit River in this c. 1938 photograph. The FF-2 was a dual-control trainer version of the FF-1 fighter and the first Navy fighter with retractable landing gear. Twenty-three FF-2s were converted from retired FF-1s and used by Reserve squadrons. Compared to other Grosse Ile aircraft, they were powerful machines.

Crewman Oscar Phillips works on a Grumman FF-2 in the Seaplane Hangar in this c. 1936 photograph. During the Great Depression, money for reservists was hard to come by. On May 14, 1933, the Navy cut reservists' pay in half, followed by a second substantial pay cut a week later; weekend reservists received no pay at all. No supply officer was assigned to Grosse Ile until 1940. Instead, a local civilian named Benjamin Micou supervised the base supplies without pay. Grosse Ile's churches supplied clergymen to fill the duties of a base chaplain. During the Depression, the Works Progress Administration (WPA) initiated a $250,000 state work relief project, adding 50-foot-wide paved runways, warm-up aprons around the landing field, and a rifle range near the blimp hangar. In 1935, another WPA project added a 10,000-gallon gasoline system.

Vought Navy Plane -Wright "Whirlwind"

This photograph shows a pair of Vought UO-1 observation planes at Grosse Ile. Like many aircraft assigned to Grosse Ile, the UO-1 was already obsolete when they arrived. Some versions were converted to fighters simply by covering over their front cockpits. As the caption indicates, the plane was powered by Wright J-4 Whirlwind engines. Wright Whirlwinds also powered Lindbergh's *Spirit of St. Louis* and Grosse Ile's ZMC-2 blimp.

Another Vought aircraft based at Grosse Ile was the SB2U-1 Vindicator scout bomber. A decade separated the SB2U-1 from the earlier UO-1 pictured at the top of the page. The Vindicator was one of the Navy's first monoplanes and the first monoplane based at Grosse Ile. Though it looks modern, it was still covered by fabric. At 230 miles per hour, it was little faster than the base's Grumman FF-2 biplanes.

An airman in the 1930s prepares for gunnery training. Enlisted men ate their meals alongside officers in the old flight school barracks, but they lived in barracks at the lower base. In 1936, a WPA project finally brought city water to the island. In 1938, the base's original Torpedo Squadron VN-9 was re-designated VS-8R, while Marine Squadron VO-5MR was replaced with VNS-5R.

The FF-2 was the last fighter to be based at Grosse Ile before World War II. In 1941, the base's mission changed to primary flight training. Grosse Ile exchanged its FF-2s for Naval Aircraft Factory N3N Canary and Stearman N2S Kaydet trainers. The Navy cadets were replaced by 100 British students from the Royal Air Force (RAF).

The above photograph shows the ramp and boathouse at the lower base under normal conditions, and the below photograph shows it during flooding. The lower base was created by dredging a cattail marsh, and throughout the life of the base the waters of Olds Bay regularly tried to reclaim it. After World War II, the lower base was occasionally used for beaching PBY Catalina flying boats, but mostly the lower base became home to the base's rescue boats and a large collection of personal watercraft owned by station keepers. Many trips across the river to the bars and clubs of Amherstburg, Ontario, began at this very spot.

This is the same location as it appears today. The hangar burned in December 1968. Presently, only the concrete ramp survives as a reminder that the Navy first arrived at Grosse Ile on this site in 1927. The area is currently a bird sanctuary under control of the US Fish and Wildlife Service. (Author's collection.)

Guide rails for the hangar doors are the only features still visible today. This photograph was taken from an area once located inside the hangar. Over 40 pleasure boats belonging to Navy personnel were lost when the hangar burned to the ground in a spectacular fire that is still considered suspicious. (Author's collection.)

This is the original Tin Hangar as it appears today. Originally built in Memorial Park in downtown Detroit, it was disassembled and floated down the Detroit River to Grosse Ile on a barge in 1927. Today, it occupies its second location on the base after being moved to make way for runways in 1942. The hangar is currently located in an area controlled by the US Environmental Protection Agency (EPA) and is off-limits to the general public. The EPA at one time offered the Tin Hangar to Grosse Ile Township provided they pay for its removal and construct0 a new storage building of similar size in its place. Another possible destination for this historic hangar is Greenfield Village in nearby Dearborn. (Author's collection.)

Two

THE ZMC-2

The third partner in the development of aviation on the southern end of Grosse Ile was an innovative company known as the Aircraft Development Corporation. The corporation was the brainchild of Carl B. Fritsche, an aviation enthusiast and business promoter who teamed with record-setting balloonist Ralph Upson to create a revolutionary all-metal dirigible. Fritsche obtained backing from Henry and Edsel Ford, as well as Charles Kettering, head of research for General Motors; Alex Dow, president of Detroit Edison; and William B. Stout, a local industrialist. To build the unique airship, the group selected an undeveloped site on the southern tip of Grosse Ile. The Aircraft Development Corporation purchased 403 acres from automobile pioneer Ransom E. Olds, and they cleared a 3,000-foot-wide circle in which they built an enormous hangar that was 120 feet high, 120 feet wide, and 180 feet long. They had the expectation of lengthening it for the fleet of all-metal zeppelins they envisioned building after finishing the prototype. This was the first actual use of land on Grosse Ile for aviation purposes, preceding the flight school and the Navy base by two years.

The prototype airship was a blimp for the Navy named the ZMC-2 (Zeppelin Metal-Clad, 200,000 cubic feet in capacity). It was constructed out of Alclad, an aluminum alloy clad with a thin skin of pure aluminum. The ZMC-2 was assembled with a unique riveting device that worked like a giant sewing machine. The blimp began as two ends suspended from the hangar's high ceiling, to which consecutive strips of Alclad were riveted. As the two halves grew, they resembled upside down teacups. When both halves were finished, they were taken down and joined with a final ring of rivets.

The ZMC-2 first flew on August 19, 1929, and traveled to its permanent home at Lakehurst, New Jersey, in October 1929. It was used by the Navy for over a decade, finally being scrapped in 1941.

The ZMC-2 proved that an all-metal airship was not only possible but also practical. Unfortunately, the fleet of all-metal zeppelins envisioned by the corporation's leaders never materialized, and the corporation went bankrupt in 1932, another victim of the Great Depression.

This c. 1929 photograph was taken from a biplane looking southwest at the blimp hangar (area "E") where the ZMC-2 was built. Meridian Circle Road is visible looping to the right. The Navy's Tin Hangar (area "F") is visible near the bottom. The access road next to it runs north joining with East River Road near the bridge to Elba Island. The circles next to the blimp hangar mark the ZMC-2's mooring mast. (Courtesy National Archives.)

This photograph is looking southeast over the southern tip of Grosse Ile with Elba Island on top. The blimp hangar is surrounded by trees in this photograph; the trees would be gone by the late 1930s. The Navy Reserve Air Station hangar and buildings are clearly visible on Olds Bay just above the blimp hangar. Frenchman's Creek is visible looping through the bottom right corner.

Construction of the world's first all-metal airship drew many aviation celebrities to Grosse Ile. Among the famous pilots who visited Grosse Ile to view its construction were Charles A. Lindbergh and Amelia Earhart. Here, a rare Curtiss Racer is seen at Grosse Ile Airfield in this late 1920s photograph, possibly belonging to a visitor who came to see the wondrous airship.

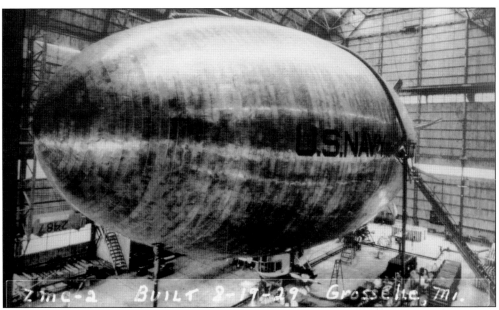

The newly completed ZMC-2 is seen stored in its hangar in this c. 1929 photograph. The narrow bands of Alclad that joined to create the blimp are clearly visible. The control car could seat four, with additional space for two observers. The engines are two Wright J-5 Whirlwinds, which were mounted in a forward-facing (tractor) arrangement instead of pusher style, as commonly used on Navy blimps.

Filling the ZMC-2 with Helium turned out to be more difficult than expected. Because helium mixes freely with air, it was necessary to remove all the air from the ship before adding the helium. This was accomplished by first filling it with CO_2. The large tube visible emerging from the blimp's lower front is for venting the CO_2 as helium was added from the top. Shortly before filling began, a bright young engineer noted that once filled with CO_2, the ZMC-2 would weigh many thousands of pounds more than when filled with air. The filling process had to be delayed several weeks while additional reinforcement cables and support panels were added to keep the ship from tearing apart.

On August 19, 1929, the ZMC-2 emerged from its hangar for the first time. The launch attracted a small crowd of Grosse Ile residents, who had waited for four years for this moment. The massive size of the ship's hangar is clearly visible as well. It remains the largest structure ever built on Grosse Ile. The ring of fins around the airship's tail is a distinguishing feature of the ZMC-2.

The ZMC-2 remained at Grosse Ile undergoing test flights until October 1929. During testing, it was discovered that the amount of helium lost to leakage was far less than in a conventional fabric blimp. Throughout its operational life, ZMC-2 required far less helium replacement than any other airship. It also had a tendency to "burp" and "pop" as the metal expanded and contracted due to temperature changes.

The ZMC-2 took its first full test flight on August 26, 1929. Note the groups of ground handlers on either side holding down the airship with ropes (below). Throughout its life, the ZMC-2 was never painted, remaining bare aluminum with the words "US Navy" and "ZMC-2" displayed on its surface. Affectionately called the "Tin Bubble" by its crews, it must have been a splendid sight with the sun reflecting off its metal surface. In October 1929, the airship finally left Grosse Ile for its new home at Naval Air Station Lakehurst in New Jersey. On the way, it stopped to attend the 1929 Cleveland Air Races, where it was seen by thousands of aviation enthusiasts. It arrived in New Jersey ahead of schedule, dodging storms along the way, where it was formally accepted by the US Navy. Unfortunately, the Aircraft Development Corporation ran into design problems with the ZMC-2's successor, and the Tin Bubble would remain the only all-metal airship ever built.

This incredible photograph shows the ZMC-2 in a hangar at Lakehurst, New Jersey, dwarfed by the German zeppelin *Hindenburg*. The ZMC-2 measured 148 feet, five inches long and 52 feet, eight inches wide. In contrast, the *Hindenburg* was 803 feet, 10 inches long and 135 feet wide. The *Hindenburg* was so large the ZMC-2 could have fit inside it sideways! The ZMC-2 was never used operationally, remaining an experimental proof-of-concept prototype. By 1936, the airship had traveled over 80,000 miles. In its lifetime, the ZMC-2 logged 752 flights with 2,265 hours of flight time. Unlike the *Hindenburg*, its flights were virtually trouble-free. The only significant accident occurred shortly after its arrival in New Jersey, when the ZMC-2 suffered a hard landing that pushed its control car over a foot up into the envelope. Since the airship's helium could not be removed, fixing the damage was no easy task. After nearly suffocating while surveying the damage, one of the original builders was fitted with an airtight suit in which he was able to enter the helium-filled envelope and affect the repairs. (Courtesy National Archives.)

In a fall 1960 photograph (above), the blimp hangar is seen being razed. For over 35 years, it was the largest building on Grosse Ile and remains the largest structure ever constructed on the island. Since the early 1950s, it had been painted bright red and white checkerboard to be visible to aircraft landing at the base. By 1960, it was in poor condition and home to a flock of Starlings that were endangering aircraft. Worse yet, it was located in the path of incoming aircraft. Eventually it was torn down. Interestingly, the roof of the rear extension, built in the 1930s, was carefully removed and trucked to Trenton where it was reused in the building of a bowling alley on West Road (below), no doubt the only bowling alley ever constructed out of a blimp hangar.

Three

THE WAR YEARS

By January 1941, all Navy Reserve squadrons were called up, and the training of new cadets was accelerated. The mobilization called for 2,905 aviators in the first three months alone. Previously, Grosse Ile had been training only 23 students a month. In August, 100 British aviators arrived to train for the RAF, and the American cadets were transferred. By the war's end, over 1,800 British pilots had trained at Grosse Ile—a record for any American base. Eleven men who lost their lives are buried at Oak Ridge Cemetery.

In 1942, American pilots returned after a $5 million construction program began, nearly doubling the size of the base to 604 acres. In only a few weeks, several buildings were constructed north of Groh Road. In addition to barracks, the buildings also housed classrooms, a medical center, gymnasium, post exchange, and a Link Trainer room. Even an Olympic-size swimming pool was installed. Hangar One was built next to the old Curtiss-Wright hangar and would house the base's administration offices. Further east, a huge drill hall with an arched roof was constructed.

In order to accommodate modern aircraft, three runways, each 150 feet wide, were laid out in a triangular pattern around the former Meridian Circle Road. The longest one, Runway 3-21, was nearly 5,000 feet long, extending from the northeast corner of the base to Olds Bay. To the west, Runway 17-35 extended 4,580 feet alongside Frenchman's Creek. Runway 9-27 ran along the front of the hangars and was 3,850 feet. Previous trainer aircraft based at Grosse Ile were replaced with the Boeing N2S-5 Stearman biplane. Sixteen airfields throughout the downriver area were obtained by the base for short runway training and emergency use.

By mid-1942, there were over 800 naval aviation cadets in training at Grosse Ile. Incoming Navy personnel more than doubled the island's population, creating a critical food and housing shortage. Island and downriver residents began renting out bedrooms to cadets. Meanwhile, for entertainment, the Olds Mansion on Elba Island became the USO Club, and cadets made unofficial visits across the river to Amherstburg, Ontario.

The flood of Navy personnel subsided in 1945. Due to the quality of training at bases like Grosse Ile, aerial combat in the Pacific resulted in fewer losses than anticipated. Grosse Ile had done its job.

The arrival of World War II would transform the base. On December 9, 1942, its name officially changed to US Naval Air Station Grosse Ile in recognition that the base had become the Navy's largest primary flight training facility. In this photograph, the 50-foot-wide runways constructed by the WPA in the early 1930s have been widened to 150 feet, obscuring the original circular landing field and Meridian Circle Road. In 1941, the Michigan Legislature arranged for the Navy to acquire the property on which the base was located from the Detroit Aircraft Corporation for $600,000. Overnight, the base grew from 375 to 604 acres. The biggest changes would occur on the north side of Groh Road, where a $5 million construction program began. In the span of a few weeks, over a dozen wooden barracks sprouted up where woods and farmlands had been. Additional buildings would go up south of Groh Road, including a new barracks next to the Officers' Club and a huge drill hall east of Hangar Two.

The first new construction was a hangar with administration offices that were located a few hundred feet west of the former flight school. The area is photographed here on June 24, 1941. It was finished in early 1942. The new hangar on Groh Road meant aircraft no longer needed to be serviced at the lower base. After 1942, that location was used exclusively for the station's various watercraft, including a small fleet of rescue boats.

At the time the new hangar was being built, the only structures on Groh Road were the former flight school hangar and its barracks (lower left). Built for the flight school, the barracks had housed the base's station keepers since 1932. Behind Hangar Two, ground has been cleared for construction of the base's new steam plant.

Once finished, the new hangar became known as Hangar One. In these photographs, work has just begun on the west administration wing and its observation/meteorology deck. Additional offices and maintenance shops were located on the east side. The woods and farmland on the north side of Groh Road would soon be cleared for construction of barracks and halls. The original Navy barracks would become the Officers' Club. Just west of the Officers' Club, construction has begun on a building known as the White Barracks. Note that the control tower was not part of the original construction. Until the mid 1940s, the base used a smaller control tower located on top of Hangar Two.

Hangar One nears completion in this photograph taken on October 15, 1941. The hangar was built before the base's runways were widened to 150 feet. On the extreme left, the steam plant and water tank located behind Hangar Two are finished. In the upper right is the Navy's original Tin Hangar, which would be moved to accommodate runways in 1942. In the lower left, the foundation is laid for the White Barracks. A group of Yellow Pearl or Kaydet biplanes are visible on the ramp and field in front of Hangar Two.

Here is Hangar One in its finished form. The control tower was added in the mid-1940s. To the left of the Officers' Club are the White Barracks, which have been completed. Across the street are row upon row of new wooden barracks, built on newly paved roads named for aircraft carriers, such as *Hornet* and *Yorktown*. Midway Road runs north to south, connecting the base. The small white farmhouse visible just above the White Barracks would remain throughout the base's existence. One of its former residents remains active in the NASGI online community. At the top is Detroit Edison's enormous Trenton power plant, commonly known as the "Trenton Stacks."

The steam plant under construction is shown above on October 15, 1941. It is shown below completed on December 15, 1941. Located on Groh Road just behind Hangar Two, the steam plant provided heat to most of the base's buildings. Next to it, a large water tank was added at the same time. Though the steam plant was the second new structure started on the base, it was the first to be finished. Unused since the base closed in 1969, the building was finally razed in the 1990s. In the background are the base's two original radio transmission towers; both were taken down by the war's end.

The last building to go up north of Groh Road was the base's recreation hall. The hall's auditorium is being completed in this photograph taken on November 2, 1942. The stage area is visible in the rear. Overhead are the hall's massive wood beams, more than three feet wide and a foot thick. Veterans wonder why these beautiful beams were not salvaged when the building was razed in the 1990s.

Photographed on November 12, 1941, the only barracks building added on the south side of Groh Road was known as the White Barracks. It would house the base's WAVES (Women Accepted for Volunteer Emergency Service). Unlike their Army counterparts, WAVES were official members of the Navy. During the war, many WAVES achieved the rank of chief and some became officers. These women served only stateside, although some were stationed on Hawaii late in the war.

Construction of barracks and facilities north of Groh Road occurred at lightning speed, as the base prepared for the rush of Navy aviation cadets expected to soon arrive. The photograph above was taken on March 19, 1942, and the image below was taken on April 15, 1942. In 1940, the base was training an average of 23 Reservists a month. By 1944, the number of active-duty personnel would be over 2,000. In 1935, the base had only about 34 station keepers; in 1943, there were hundreds. The arrival of that many Navy personnel more than doubled the population of Grosse Ile, and still more were coming.

The photograph above, taken on May 22, 1942, shows the ground cleared for the recreation hall. It was the last structure built north of Groh Road and also the largest. The image below shows that the recreation hall was a tall, stately building, which served as the center of life on the base. The recreation hall contained the base post exchange, library, pool hall, gymnasium, and a bowling alley. The backbone of Grosse Ile was Midway Road, which extended from the front doors of Hangar One straight to the front steps of the recreation hall. Midway was used as a parade route, as well as for inspections and concerts.

This c. 1942 photograph shows the completed Hangar One. In the upper left is Hangar Two, adjacent to the steam plant and water tower. The old radio towers have already come down. The letters "USNR" on the roof of Hangar Two indicate that the image predates the base's name change. Pictured in the center is the newly completed Hangar One with its administration wing and observation tower; it still lacked the control tower. A dozen Yellow Pearl and Kaydet trainers are parked around the ramps. In the lower right is the Officers' Club. All three of these buildings still exist today.

The massive drill hall dominates the eastern end of the base in this photograph that also shows virtually every building on the north side of Groh Road. When the number of personnel assigned to the base dropped substantially after the war, the drill hall was used as a hangar for servicing aircraft. Leaking fuel and lubricants soon destroyed the polished wood floors on which thousands of servicemen had spent countless hours drilling. Today, the restored hall is home to the Grosse Ile Tennis Courts. In the lower right are two of the base's earth-covered ordnance bunkers. The small building just across the street from the hall is the Airport Inn, a small restaurant located just outside the base's main gate. It still exists, continuing to serve hungry pilots and tennis players.

The Officers' Club and BOQ are located on Groh Road in front of Hangar One. Built in 1927 as a barracks for the Curtiss-Wright Flight School, it was enlarged in 1938. When the base closed in 1969, the Officers' Club was among the first buildings leased by the township for private use. The building's huge upstairs social hall became a popular restaurant known simply as the Officers' Club. The restaurant was a success, drawing crowds of islanders, as well as residents of Trenton, Wyandotte, Gibraltar, and other downriver towns. Since then, it has been home to several businesses and is currently a luxury restaurant and hotel known as the Pilot House.

In 1941, Grosse Ile traded its reservists for 100 British RAF flight students. With war raging in England, and British facilities and aircraft stretched to the maximum, the RAF began shipping flight students across "The Pond" for training in the United States. Grosse Ile was one of the first bases to receive British students. Here, a group of RAF students and their kit bags are unloaded at the base. They arrived without uniforms, having only an RAF badge, black necktie, and British accent to distinguish them from their American counterparts. Training consisted of one month of ground schooling before checkout flights in the base's Stearman Kaydets. Those who passed moved on to additional training in Pensacola, Florida, before returning to England as pilots. Many would serve in the Fleet Air Arm, flying Fairey Swordfish, a biplane very similar to the Stearman.

Two graduation class photographs show RAF pilots trained at Grosse Ile in 1942–1943. The image above from 1942 shows an early class of 23 students, the same class size the Navy Reserve had trained since the late 1930s. By 1943, growing pressure to produce more pilots increased class sizes by 10 or more students, as seen in the class photograph below. This image is also interesting for the wide variety of flight clothing worn by the students, including pre-war vintage Navy flight coveralls likely passed from one student to the next since they were issued in the 1930s. Several students, including a tall fellow (back row, far right), are wearing the Navy's new ANJ-4 sheepskin flight jacket. Curiously, two men (back row, far left and front row, far right) are wearing Army Air Corps A-2 flight jackets. All of them are wearing US-issued sheepskin flying boots.

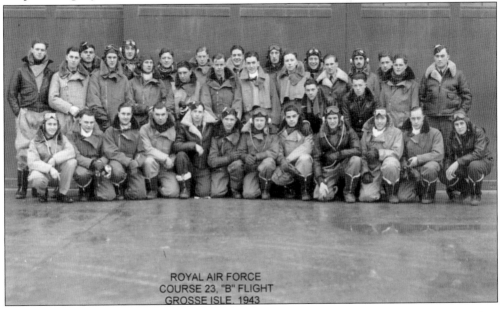

ROYAL AIR FORCE
COURSE 23, "B" FLIGHT
GROSSE ISLE, 1943

British officers pose for a photograph before shipping back to England. Over the span of four years, Grosse Ile trained more British pilots than any other US base. The training the British students received was a crash course to weed out those who would not qualify as pilots. As a result, Grosse Ile earned a reputation as an "E base," or elimination base.

This photograph shows the RAF versus the US Navy in a muddy game of soccer. The British students were well liked by both base personnel and Grosse Ile residents, and many found housing in local homes during their brief stay on the island. The rate of training accidents for British students was below that for their American counterparts.

In the c. 1942 photograph above, British airmen enjoy a party in their honor at the Grosse Ile Country Club. Below, British Cadets enjoy the hospitality of the USO at the former Olds Mansion on Elba Island. Many men were away from home for the first time, and Grosse Ile residents went out of their way to make them feel comfortable. There is no doubt this had a strong impact on the high amount of passing grades awarded to RAF flight students. As one cadet said, "The good part of our stay at Grosse Ile was the wonderful hospitality provided through the USO. Any serviceman could register and say how he wanted to spend his time, and a local family was produced who would take him to whatever he wanted in the form of entertainment."

British airmen pose for a photograph, most likely at the Olds Mansion's USO Club. The residents of Grosse Ile can be proud of the contribution they made to the war effort. Even before war was declared, British flight students were arriving on the island. The last one departed in April 1944. During that time, more than 1,800 British airmen received training at Grosse Ile.

Eleven British airmen who lost their lives while training at Grosse Ile are buried at Oak Ridge Cemetery near the intersection of Telegraph and West Roads. On a Sunday every May, the Royal Canadian Legion conducts a memorial service at the cemetery. In 1958, the bodies of six RAF pilots killed in a Vulcan bomber crash in Detroit were buried alongside.

In 1942, after a year's absence, US Navy cadets returned to US Naval Air Reserve Base Grosse Ile. By fall of that year, the base was home to over 2,000 officers, enlisted men, and British flight students, making it the Navy's largest primary flight school. In recognition of its status, the Navy changed the base's name on December 9, 1942, to Naval Air Station Grosse Ile. The base's Grumman FF-2s and Curtiss Helldivers were gone, replaced by row upon row of N3N-3 Yellow Pearl and Boeing Stearman Kaydet biplane trainers. By 1944, the later N2S-3 version of the Stearman was the only airplane in use at NASGI.

The 40th pilot class poses in front of a Boeing N2S-1 Stearman Kaydet in November 1942. This class was made up of British Royal Navy pilots bound for service flying torpedo bombers and scout planes from the decks of British aircraft carriers. Like their US Navy counterparts, the British Fleet Air Arm received training at Naval Air Station Grosse Ile and played a significant part in the war's victory. Unlike their American counterparts, British Navy pilots fought almost entirely against German foes. Combat operations involving the Fleet Air Arm resulted in the sinking of the German battleships *Bismark* and *Tripitz*, as well as countless submarines and cargo ships. Later in the war, Fleet Air Arm pilots would acquire Vought F4U Corsairs, the same aircraft flown by their American Navy counterparts.

Flight crews ready a N3N-3 Yellow Pearl for a training flight. During the early 1940s, the N3N-3s shared the flight line with Boeing N2S-1 Stearman Kaydets at Grosse Ile. Produced by the Naval Aircraft Factory in Philadelphia, Pennsylvania, the N3N was entirely designed and produced by the Navy without assistance from civilian companies. Even the plane's Wright R-760 Whirlwind engine was produced under license by the Navy. Over 1,600 aircraft had been manufactured by the time production ended in January 1942. N3N-3s were the last biplanes used by any branch of the US military. The last one was retired at the US Naval Academy in 1961. It is also the airplane seen chasing Cary Grant in the film *North by Northwest*.

The Wing Shop was a part of the base's Assembly and Repair Unit. Here, skilled seamstresses recover the wings of Stearman biplanes. After this, the wings would go to the Dope and Paint Shop before being reattached to an airplane, though not necessarily the same plane they were removed from. Assembly and Repair was responsible for all aircraft maintenance and had several departments dealing with specific aspects of aircraft repair and service. With a fleet of fabric-covered biplanes to maintain, the Wing Shop was constantly busy. Most of the women, who were trained in the art of recovering fabric aircraft, were not from the Navy but instead the Grosse Ile and downriver area.

The duty of the Assembly and Repair Unit was to maintain, repair, and overhaul aircraft and engines. Here, two members work on a Boeing Stearman Kaydet. The unit played a vital role in making sure the base's aircraft were safe to fly. Student pilots had enough to worry about without having to be concerned about the plane falling apart around them.

Newly arriving Navy cadets are seen here getting a quick medical inspection. Such inspections identified any communicable diseases the cadets may have brought with them, and they were conducted in an assembly line fashion. The base had a well-equipped medical center located among the barracks north of Groh Road. It could handle most illnesses and minor injuries. Serious injuries were generally treated at hospitals on the mainland.

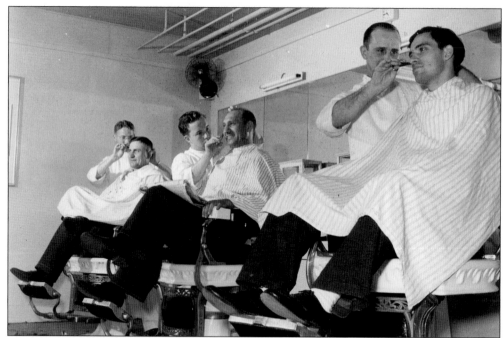

The base's service shops had all the essential services needed to keep a Navy man looking good. Most shops were located among the barracks north of Groh Road. They employed enlisted personnel who served as tailors, cobblers, and barbers. These men and women were knowledgeable in the Navy's specifications for uniform, dress, and grooming. With over 2,000 people on base, a large full-time staff was required. After the war, the staff was cut back dramatically. Only one barber was on staff throughout the 1950s. Sadly, the service shops were often the last facilities on the base to be updated. In the 1960s, much of the base's sewing and pressing work was still being done with the same equipment used during World War II. All the services were rendered for free, though tipping for a particularly good haircut was not discouraged.

A cadet's first two weeks of flight training were spent entirely in the classroom. By this point, the cadet had already gone through basic training and a comprehensive medical and written exam before being selected for primary flight training. Over two-thirds of those showing an interest in flying carrier aircraft never got this far; those who did were by no means guaranteed a future as a pilot. Grosse Ile was known as an "E base," or elimination base. Its job was to find those cadets who were not going to make it through flight training and get them out early, sending only the best on to finish their training at Naval Air Station Pensacola. The image below shows the base's radio ground school.

With 2,000 hungry mouths to feed, the commissary department had its hands full. No sooner was one meal served than another was being prepared. Over the years, Grosse Ile had many cooks on its staff—some were very good, while others were not. Many veterans remember the food at NASGI as being well above average compared to other Navy bases. The base's kitchen and mess hall was a large, open building located among the barracks north of Groh Road. It was constructed in 1941. Prior to that, Grosse Ile's reservists ate in the old flying school barracks that later became the Officers' Club and BOQ.

This image was taken in the drill hall on May 1, 1942. At that time, the civilian population of Grosse Ile was around 2,000—the base population was over 2,000. Even the addition of 225 acres of new housing north of Groh Road could not eliminate the housing problem. Soon, residents of Grosse Ile and other downriver communities began renting out their spare rooms to Navy men and British flight students. As the community's young men began heading off to war, it was often their bedrooms that were offered to NASGI personnel, as families exchanged a son for an airman. Food became another problem. The invasion of 2,000 Navy personnel overloaded Grosse Ile and Trenton's grocery stores, leaving the civilian population searching downriver communities for food at a time when rationing was already impacting supplies. Grosse Ile's food shortage was not solved until 1944, when a well-stocked store was opened on the base.

In addition to aircraft, Hangar One also had the base's pistol range. While at NASGI, aviation cadets were expected to learn the basics of flight theory, air-to-air combat, dive bombing, torpedo bombing, aerial scouting, formation flying, navigation, radio communication, aircraft maintenance, aerial gunnery, escaping from a ditched aircraft, parachuting from an airborne aircraft, packing a parachute, basic first aid, and, as seen above, how to shoot with a pistol. They had one month to convince their instructors they were qualified to be passed on to more advanced training or they were sent back to the Navy for training in a different field.

Here, students are training on an air-to-air combat simulator. Oncoming enemy aircraft were projected onto a screen, and the student had to use the device's machine-gun sights to track and hit them. This photograph shows the degree of sophistication that cadets found when they arrived at Grosse Ile. Little expense was spared while preparing students to fly the Navy's newest and best combat aircraft.

The base library, shown here around 1944, was located north of Groh Road in the recreation hall, and it was well stocked. Navy manuals and reference books, as well as popular literature, were found on its shelves. Veterans recall that current best sellers were often available at the library only days after their publication.

No military training is ever complete without drilling. This photograph was taken on February 13, 1943, on what must have been an uncharacteristically warm February day in southeastern Michigan. The cadets are drilling along Midway Road. The road was aptly named, as it ran midway through the base and honored the Battle of Midway, fought just before it was constructed. The Officers' Club is visible in the background.

The Link Trainer was an innovative device. Created by Ed Link in 1929, it was an enclosed metal cockpit mounted on a pedestal moved by bellows that could pitch and bank. It reflected the aircraft's movements on a simulated instrument panel inside. During the war, over 900 of these devices were produced to train Army and Navy pilots.

Seen here is the moment every flight student waited for. After weeks of classroom instruction, medical exams, and even pistol training, the cadets finally received their first flights. The first order of business was dressing for the occasion. Even in summertime, flying in an open cockpit at 10,000 feet in the sky was guaranteed to be cold. Each cadet wears an early style Navy M-422 sheepskin flight jacket and sheepskin-lined flying boots, as well as their "Mae West" inflatable life preservers. The next step was selecting a parachute, as seen in the above photograph. It was then on to the mission boards, seen below, which told cadets who would be their instructor, or if they were an instructor, which cadet was theirs to terrorize for the day.

The board showed flight times, the instructor, and the students, as well as the type of training the student was to receive. Flight training went on for several weeks, during which time the student was taught to takeoff and land, as well as basic maneuvering of the trainer. As shown below, flight instructors were seasoned pilots often with combat experience. They were constantly on the lookout for students who were uncomfortable, uncoordinated, or just plain reckless so that they could eliminate them early. Once finished at Gross Ile, those who moved on to Naval Air Station Pensacola still had a long way to go before earning their wings, but at least the Navy knew they were investing precious flight training upon only the best students.

A training class pauses for this c. 1943 group photograph in front of the mission boards. This was likely a posed shot, since none of the students are wearing their flying boots or parachutes and their Navy "Dixie Cup" hats would have been useless in a cockpit. During the war, NASGI trained hundreds of classes like this one. During their month at the base, each student was closely supervised, their abilities assessed, and a decision made to pass or fail them—there was no second chance. If a student was deemed unsuitable for pilot training, they were sent elsewhere for training in another field. However, some students were able to qualify as radio operators or navigators and remained in aviation.

Naval Air Station Grosse Ile had several celebrities appear on its recreation hall stage, but it also had two future celebrities from its World War II–era ranks. For two months in the spring of 1945, a young Lt. George Herbert Walker Bush was assigned to temporary duty with Torpedo Squadron 97. Newly married Lieutenant Bush and his wife, Barbara, lived in a home in nearby Trenton during their time at the base. From August to October 1945, future game show host Lt. (JG) Bob Barker was also based at Grosse Ile. Of his time there, this is what he jokingly had to say, "Immediately upon arriving at Grosse Ile, I checked out in the Corsair, and I was assigned to a fighter pool awaiting orders to join a seagoing squadron. When the enemy heard I was headed for the Pacific, they surrendered. That was the end of World War II. I was discharged and returned home, a hero in my mother's eyes!"

When constructed in 1942, the station's Olympic-size swimming pool was the third largest in the state. It was still the third largest when it was torn up in the 1990s. The pool was used for recreation and water shows, as well as to train pilots how to escape from a submerged cockpit. Diving from the rafters was a popular, though hazardous, stunt.

A popular gathering place was the Ship's Service Canteen, located in the recreation hall. The canteen served snacks and coffee, and it was also home to the base's display cases. The cases contained awards and trophies won by the base and its servicemen, as well as various Navy artifacts. The walls of the canteen were covered with large oil paintings depicting modern warships and recent sea battles.

A small bowling alley was located in the basement of the recreation hall. There was also a gymnasium, library, auditorium, and pool hall located in the building. These photographs from 1942 show the base's bowling team, which competed regularly against civilian teams throughout the downriver area. The base also supported a baseball and basketball team.

Like most large military bases, Naval Air Station Grosse Ile had its own band. The c. 1943 image above shows the band during a daily concert on Midway Road. The photograph below depicts it posing in front of the base gazebo. The band gave concerts daily for the entertainment of base personal, and it also performed in parades in Detroit and the downriver area. Beginning in 1942, it competed nationally against other military bands. The band was one of the first base specialties to be interracial; membership was based upon ability, not rank or color. Band members followed a busy rehearsal schedule and were exempt from some of the base's less desirable duties. Note the decorative aircraft art on the band's bass drum.

The recreation hall's auditorium featured a stage and dance floor. Throughout the 1940s, it was the site of countless dances, musical shows, and celebrity appearances—even Ed Sullivan once took the stage here. Above, the Coca-Cola "Parade of Spotlight Bands" featured Jan Savitt and his orchestra in a broadcast from the auditorium on April 3, 1943. The c. 1944 photograph below shows a rehearsal of the base's mixed chorus. The chorus performed both popular songs as well as religious hymns, much of it arranged by popular bandleader Fred Warring. Warring was sometimes referred to as "the man who taught America how to sing."

For 25 years, the recreation hall was the site of countless base dances. Pictured above is the base's Halloween "Spook Dance" in 1943. Dances were usually sponsored by the USO Club on Elba Island and attended by local young ladies who were also glad to have time away from the war effort. After the war, servicewomen based at Grosse Ile provided dance partners, but during the war years it was the local girls who made so much difference in young airmen's lives. Over the years, a good many romances and more than a few marriages had their beginnings at these popular Grosse Ile dances. When not being used as a music or dance hall, the auditorium also served as the base's basketball court. The court's markings are visible on the hall's hardwood floor.

The recreation hall was also the site of holiday events for base families. Pictured here is the base Christmas party in 1944. As the war wound down, there were less students arriving at NASGI, and the base became home to returning fighter and torpedo bomber squadrons. Many of the base's personnel were arriving with wives and children who were housed in the married men's quarters or off the base. As a result, the base developed activities and events to help military fathers entertain their families. The annual base Christmas party was an example of the way the Navy supported it youngest members. Here, a jovial, skinny Santa Claus passes out gifts to base children, many of them dressed in scaled-down versions of their father's uniforms. The recreation hall is remembered affectionately by base veterans. It was left in good condition when the base closed in 1969, and veterans expected it would continue to be used by the township as the town's recreation hall or school; instead, it remained vacant until being razed in the 1990s. The recreation hall's destruction still draws strong feelings from NASGI veterans.

This c. 1943 photograph shows the Olds Mansion on Elba Island. It was built by automobile pioneer Ransom E. Olds in 1916 and sold to the Aircraft Development Corporation in 1925. Carl B. Frisch turned the mansion into the Chateau Voyageur, a home for the pioneer air club he had founded. The Voyageur's members were active throughout the 1920s, participating in numerous flying events. Their specialty was long-distance balloon travel. When the mansion was leased to the Navy in 1933, it was left vacant until the rising tides of war in 1942 transformed it into the island's USO Club. The USO sponsored dances, parties, and community events in support of the island's local servicemen. First to benefit from the USO's generosity were the hundreds of British airmen sent to Grosse Ile in 1941. They paved the way for thousands of Navy airmen who would follow. The 12,000-square-foot house has about 30 rooms. With a spacious front yard, expansive rear lawn, and a sitting area on the water's edge, it could accommodate very large groups. Today, the mansion has been turned into an elegant, if somewhat eclectic, apartment building.

Above is a c. 1948 photograph of the Olds Mansion, and below is an image of how the mansion looks today. When finished, it was the largest home in Grosse Ile Township (though located on Elba Island). It remained the area's largest residence until the proliferation of mega-mansions in modern times. The image above was taken from the circular stone sitting area (see facing page). In the modern photograph, it is clear that the house has changed little over the last half century. Walking around the spacious back lawn, one can easily imagine a young airman stepping from the house at any moment or the sound of laughter and music coming from its ballroom. It is a haunting reminder of a bygone era. (Above, courtesy GIHS; below, author's collection.)

The circular stone sitting area is shown both in 1943 (above) and in 2011 (below). Over the years, this site has entertained automobile inventors, aviation pioneers, British airmen, Navy pilots, and at least one future president of the United States. Visible across the mile-wide Detroit River is the Ontario town of Amherstburg. Unofficial trips across the river to the clubs and bars of Amherstburg were a popular distraction for Navy men for over 40 years. In the late 1940s, this section of river was a common place for unauthorized air races in Navy Hellcats and Corsairs. More than one Navy plane came to rest on the river bottom along this stretch. In recent years, there have been attempts by salvage groups and historians to locate some of the aircraft on the river bottom; none have been found. (Above, courtesy GIHS; below, author's collection.)

The photograph above shows a group of airmen and Navy WAVES enjoying time at the USO Club on Elba Island. The image below depicts the same location in 2011. The view across the Detroit River offered visitors an opportunity to see the many cargo vessels that carried the products of Detroit's war effort to the men on the front lines. Among the sights these sailors might have seen were transport ships loaded with Sherman tanks from the Detroit Tank Arsenal, as well as countless coal freighters bound for the steel refineries that once lined the Detroit River from Trenton all the way to River Rouge. Coal-fueled refineries turned out steel that traveled only a few miles before being turned into tanks, trucks, Jeeps, and airplanes. Note the same tree appears in the background in both photographs. (Above, courtesy GIHS; below, author's collection.)

During the war, the Olds Mansion's USO Club was a retreat from the business of war. The photograph above was taken at the club's annual Christmas Eve party in 1942, while the image below, taken on October 25, 1944, shows a gathering of airmen enjoying a song with USO volunteers. By this time there were 3,000 USO Clubs operating in the United States. The USO was founded in 1941 in response to a request from Pres. Franklin D. Roosevelt, who wanted a place for servicemen to go where they could feel "at home" while on leave. This request brought together the Salvation Army, Young Men's Christian Association (YMCA), Young Women's Christian Association (YWCA), National Catholic Community Service, National Travelers Aid Association, and the National Jewish Welfare Board, forming a single organization that provided entertainment and moral support for hundreds of thousands of servicemen during World War II and millions more since. Its second chairman was Prescott Bush, whose son, future president George H.W. Bush, would enjoy its benefits at the USO Club on Elba Island.

Above, the first group of Navy WAVES to arrive on Grosse Ile is pictured alongside a North American SNJ trainer in May 1943. Below, the WAVES are photographed with skating superstar Sonja Henie. The WAVES organization was formed in August 1942, two months after the WAAC (Women's Auxiliary Army Corps) was established. The major difference between the two was that the WAAC was an auxiliary organization that served alongside the Army, while the WAVES was a full-membership organization of the Navy; WAVES members were capable of becoming commissioned officers. During their stay on Grosse Ile, WAVES were housed in the White Barracks located next to the Officers' Club. They were barred from service outside the United States, though some served in Hawaii late in the war.

This c. 1943 photograph shows a Grosse Ile honor guard on parade in downtown Detroit. Naval Air Station Grosse Ile was the largest Navy base in Michigan and played a visible role throughout the Detroit area. During World War II, formations of Navy Stearman and Yellow Pearl trainers were a constant sight in the skies over Detroit. Personnel stationed at Grosse Ile could be seen at Detroit Tigers baseball games, Lions football games, and in 1943 watched as Syd Howe and Mud Bruneteau led the Detroit Red Wings to their third Stanley Cup by sweeping the Bruins. If Detroit was the "Arsenal of Democracy," then the men and women of Naval Air Station Grosse Ile were the daily seen faces of those who took the weapons of freedom to the enemy.

Ultimately, the reason each student came to Naval Air Station Grosse Ile was because he wanted to be a Navy pilot—he wanted to fly. The elimination process was harsh and often unfair. A student could be rejected because his hearing was bad or he was color blind. He could be turned away despite being a licensed pilot in civilian life. Sadly, potential students could be denied training if they were female or had the wrong color of skin. He could also lose his life before completing his training—and many did. For those fortunate airmen who were selected to go on, only the easy part was behind them. Ahead lay advanced training at Naval Air Station Pensacola, where they would meet a more powerful trainer aircraft, the SNJ Texan, and later such formidable machines as the Wildcat, Hellcat, Avenger, and the bent-wing Corsair. Corsairs would come to Grosse Ile in 1946 and played an important role in the next chapter in the history of Naval Air Station Grosse Ile.

Four

THE POSTWAR YEARS

Naval Air Station Gross Ile barely had time to catch its breath before entering the era of the Cold War. Naval reserve aviation officially restarted on Monday, July 1, 1946. Grosse Ile decided to celebrate a day early with an open house held on Sunday, July 30. During the previous week, a F6F Hellcat had been on display in Cadillac Park, promoting naval reserve aviation. The base began reserve operations with 10 Navy squadrons and one Marine squadron. Soon, there were 17 squadrons and 1,508 pilots flying reserve operations from NASGI: fighter squadrons were flying F6F Hellcats and FG-1D Corsairs, scout squadrons used SB2C Helldivers, torpedo squadrons had TBM Avengers, patrol squadrons began operating PBY Catalinas, and transport squadrons were flying R4Ds. All squadrons had use of the bases SNB/JRB Twin-Beech and SNJ Jaybird trainers.

In August 1949, the Navy received the last 15.1 acres of base property from the State of Michigan, finally giving them possession of the entire base. That same month, Grosse Ile's pilots achieved their carrier qualifications aboard the USS *Cabot* (CVL-28) in the Gulf of Mexico. Flying F6F, FG-1D, and TBM Avengers, 54 pilots accomplished 606 takeoffs and landings. The next best unit did barely over 200. Grosse Ile reservists performed their first Cold War mission in March 1949, when the base's mobile Ground Control Approach unit was sent to assist with the Berlin Airlift. In October 1950, two hundred men from Patrol Squadron VP-731 were sent to Korea. A fighter squadron and other units were soon to follow. By the summer of 1951, there were 12 squadrons of reserve pilots serving in Carrier Task Force 77.

In October 1951, the Navy completed a $40,000 expansion plan for the base. A construction grant of more than $750,000 was held for future use. The plan called for the addition of a longer runway on the base's east side. It would give NASGI the ability to operate the Navy's new generation of jet fighters and attack aircraft. The township's board of trustees voted it down. Unable to expand its runways, the base's fate was sealed, but "Detroit's unsinkable aircraft carrier" still had a few fights left in her.

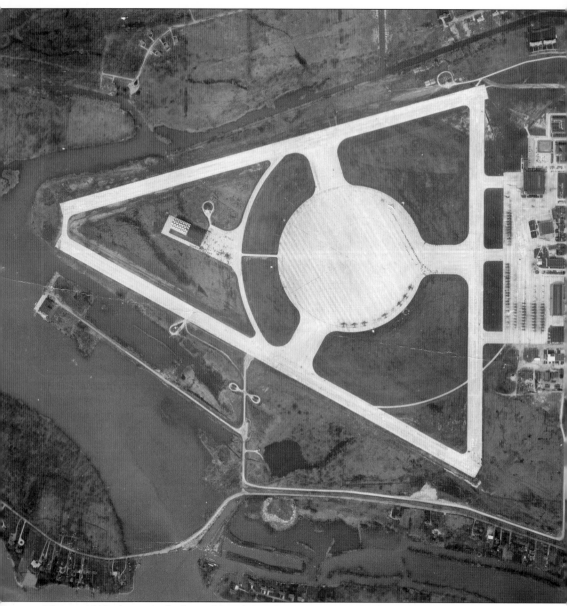

This c. 1949 photograph shows Naval Air Station Grosse Ile from the air. The base's three 150-foot-wide runways form a distinctive triangle pointing south. At nearly 5,000 feet, the longest is Runway 3-21 to the east (bottom). On the west side (top), running parallel to Frenchman's Creek, is the 4,735-foot Runway 17-35. The 3,850-foot Runway 9-27 connects them on the north. The changes brought about by World War II almost completely obscured the original base, but a few traces of the old Grosse Ile Airfield can still be found. Portions of Meridian Circle Road remain, mostly in front of the blimp hangar and in the northeast. A circular concrete pad, constructed in 1941, serves as a ramp for parking aircraft. The old quarry, where sailors used to go for a cool dip on hot summer days, is still there near the bridge to Elba Island. In a few years, the Army would be building its Nike Base between the Elba Bridge and the access road leading to the lower base. Only the seaplane hangar remains at the lower base.

One of the first new airplanes to arrive after World War II was the Consolidated PBY Catalina flying boat. Here, a late-model, radar-equipped PBY-6A takes off to scatter flowers over Lake Erie as part of a memorial ceremony on June 21, 1947. Beginning in 1946, all aircraft based at Grosse Ile had the capitol letter "I" painted on their tails. In 1956, the code would change to "7Y."

A southeastern Michigan ice storm has taken its toll on a row of parked PBY Catalinas. Ice has coated the aircraft's rear, making them tail heavy. The PBY Catalina was flown by patrol squadrons at Grosse Ile until the mid 1950s, when it was finally replaced by Grumman S2F Stoof in the antisubmarine role.

The end of World War II saw many changes come to Naval Air Station Grosse Ile, but one thing that never changed was inspections. It did not matter if one was in the Marines (above) or Navy (below), inspections were a part of daily life. The image above was taken on the ramp in front of Hangar One and shows the base's new control tower at the hangar's southeastern corner. The tower was removed in 2006. The port wing of a PBY is visible on the right. The photograph below is looking south on Midway Road. The Officers' Club is visible in the background. The flagpole still stands on the north side of Groh Road.

This is a remarkable photograph taken around 1949 from the base's new control tower on top of Hangar One. It shows a row of Grosse Ile FG-1D Corsair fighters parked on the ramp, their wings folded. Markings on the Corsair's tails contain the base's letter code "I." To the left is a row of Helldivers. Beyond the Corsairs is a row of silver North American SNJ-4 trainers. The SNJ was the Navy version of the venerable AT-6 Texan and replaced the base's Stearman biplanes in 1946 as primary trainers. Only just visible in the extreme upper right is the base's lone Beech GB-1 Staggerwing, which was used for liaison work. The sailor on the left is using the tower's signal lamp to communicate with an approaching aircraft. Though sometimes used for Morse code, at naval air stations the signal lamp usually projected a color indicating to approaching aircraft if they were cleared to land.

This photograph shows a formation of FG-1D Corsair fighters based out of Naval Air Station Grosse Ile flying over the Detroit River sometime in 1946. No aircraft personifies NASGI, or has been associated with the base's history, as much as the Corsair. The first Corsairs arrived at NASGI in 1944. They were war-weary veterans belonging to fighter squadrons returning from the Pacific. By late 1945, Corsairs were a constant presence on the base, though the first squadrons assigned to NASGI did not arrive until 1946. These aircraft are Goodyear FG-1D, built in Akron, Ohio, but otherwise identical to the F4U-1D. Grosse Ile's runways could not handle the Navy's new jet fighters, and in the spring of 1950 the base's fighter squadrons were reassigned. Fighter pilots had to transfer to attack squadrons or switch to another base. In tribute to the Corsair's legacy at Grosse Ile, the Michigan Air National Guard Museum at Selfridge displays its Corsair in the markings of NASGI.

The highlight of the year was NASGI's annual Labor Day Air Show. Planned a year in advance, the Grosse Ile air shows attracted visitors from all over the Detroit area. They were a way for the base to give back to the community for their local support and to familiarize Detroit residents with the kinds of aircraft and equipment based there. Over the years, the Grosse Ile air shows were phenomenally successful. In 1962, the base celebrated 35 years of aviation on the island with an air show featuring the Blue Angles. It attracted over 100,000 visitors, the largest attendance of any event in the island's history. Base planes are on display for visitors at the 1949 air show. Pictured are F4U Corsairs (right) and Grumman F6F Hellcats (left). In the left foreground is a Martin AM-1 Mauler. In the left background are two Douglas R4Ds (C-47). Beyond them, the blimp hangar is visible in the distance with its distinctive checkerboard paint job.

Helicopters arrived at Grosse Ile in 1959. First was the Piasecki HUP-2, a small rescue helicopter. The little HUPs quickly began earning their keep. They were regularly called upon to rescue boaters and fishermen stranded on Lake Erie. The idea of Coast Guard air-sea rescue helicopters had not evolved yet. Grosse Ile pilots were pioneering future life-saving techniques by rescuing US and Canadian citizens who had become stranded on Lake Erie. On one occasion, the little HUPs rescued 14 local fishermen who had become trapped on an ice flow during a snowstorm. In the 1960s, the HUPs were replaced by larger SH34J Seabats. Here, a Grosse Ile HUP-2 demonstrates rescuing a downed pilot during an air show at Detroit Metro Airport.

By far, the largest aircraft ever to visit Grosse Ile was the massive Lockheed R6O Constitution, seen here during its appearance at the 1949 air show. The Constitution was a double-decker transport designed during the war. It had a length of 156 feet and a wingspan of 189 feet. The maximum takeoff weight was 160,000 pounds. The Lockheed R6O Constitution (BuNo 85164) was the second of two Constitutions built. Its upper deck was furnished as a luxury passenger transport, with plush accommodations for 92 passengers and 12 crew members. A spiral staircase led to the lower deck, which had 7,373 cubic feet of cargo space. In 1949, the Constitution toured the United States, appearing at 19 cities. On its side, the words "YOUR NAVY—AIR AND SEA" were used to promote Navy recruiting. Over 546,000 people toured the aircraft's interior during its air-show visits. Before landing at Grosse Ile, the Constitution spent a half hour flying over downtown Detroit and the downriver area accompanied by a formation of NASGI Corsairs. The demonstration was two-fold, as it helped burn off fuel, which allowed the monstrous aircraft to takeoff again from the base's 4,900-foot runway.

The only jet fighter ever based at Grosse Ile was the McDonnell FH-1 Phantom. In December 1950, six of the twin-engine jets arrived on the base. The Phantom was the Navy's first jet fighter and the first jet to operate from a US aircraft carrier. The plane was something of a disappointment. Its top speed was 479 miles per hour, only slightly better than piston-powered fighters, and its straight-wing design was closer to a World War II model. Those disadvantages also meant that the Phantom could operate from Grosse Ile's short runways. The planes were never assigned to any squadron but were there to give Navy reservists a chance to check out in a jet. Unfortunately, the early jet engines were exceptionally loud and soon caused island residents to file the station's first noise complaints. Within a few months, the Navy shipped them off, but the damage was done. Shortly thereafter, the Navy approached the township with a plan to expand the base runways, making them long enough to accommodate the Navy's new generation of jet aircraft. It was voted down, sealing the fate of the base.

Reservists pose in front of a Grosse Ile–based Martin AM-1 Mauler. NASGI was one of the Navy's top operators of the massive Mauler. Designed as a heavy-lift attack aircraft, the Mauler still holds the world record for the heaviest load ever carried by a single-engine piston aircraft. Entering service in 1947, on one occasion a Martin test pilot flew one hauling: three 2,200-pound torpedoes, 12 five hundred–pound bombs, guns, and a full load of ammunition; it was a total of 12,689 pounds of ordnance, 14,179 pounds of useful load, and a gross weight of 29,332 pounds. The secret was the aircraft's 28-cylinder Pratt & Whitney R-4360 engine, the same engine that powered the six-engine B-36 bomber. Unfortunately, the R-4360 was a cantankerous beast prone to overheating. Pilots loved the AM-1's load-carrying ability, affectionately dubbing it "Able Mable." Mechanics hated the aircraft, nicknaming them "Awful Monsters." Only 151 were built, and a large percentage served at Gross Ile. NASGI received its Maulers in the spring of 1950 and was one of the last bases to retire them in late 1953. In 2010, a complete R-4360 power plant was unearthed on the outskirts of the base—no doubt discarded by a disgusted mechanic.

Failure to get permission to increase the length of Grosse Ile's runways meant that pure jet aircraft would never be able to operate there. Plans were made to close Naval Air Station Grosse Ile and move the Navy to Detroit Metro Airport or Willow Run. In the meantime, there were still aircraft that could operate from its relatively short runways. In 1957, NASGI became home to squadrons of Lockheed P2V-7 Neptune and Grumman S2F antisubmarine aircraft. The P2V-7 used two jets for extra power, making it the only other jet to operate from Grosse Ile besides the Phantoms.

Perhaps the most successful attack aircraft ever is the Douglas AD-4 Skyraider, a smaller, less complicated answer to the Mauler. It made its debut in Korea and was still in use at the end of the Vietnam War. NASGI received AD-4s in the mid-1950s. They were among the last combat aircraft assigned to the base. Grosse Ile's last Skyraiders were quietly dispatched to the West Coast in 1965, doubtless on their way to Southeast Asia.

The base's Tars and Rifles Drill Team was formed in 1958. Over the years, the precision drill team thrilled area residents with its adept handling of rifles and flashing bayonets. Formed from station keepers, the team appeared at civic, school, and sporting events and competed nationally as well. They won numerous awards.

In this photograph, the crew of a Grumman S2F reviews its final ground-controlled approach (GCA). The S2F, or Stoof, was a twin-engine plane with a four-man crew. Designed as a sub killer, Stoofs were an important aircraft at NASGI in the late 1950s and early 1960s, when the base was home to several antisubmarine squadrons.

In the photograph above, airman George Kantz stands guard at the main gate in 1963. The image below depicts the same gate as it appeared in 1965. The gate was located at the intersection of Meridian and Groh Roads. When the base closed in 1969, this was as far as most Grosse Ile residents ever ventured; beyond it lay the southernmost 600 acres of Grosse Ile.

When the gatehouse came down in 1971, the intersection of Meridian and Groh Roads was opened for the first time in 30 years. Looking south from the Airport Inn, this photograph shows where the gatehouse formerly stood. Meridian Road still connects the center of the island with the old base, but entry to the airport is less dramatic now, no longer requiring military ID or a base parking sticker. (Author's collection.)

Every military base has its watering hole; NASGI's was called the Airport Inn. Built in 1942 by an enterprising local businessman, it started out as little more than a bar and grill. Pilots quickly named it the "Down and Dirty." Located just outside the main gate, it served base personnel and their families for decades. Today, the guard shack and gate are long gone, but the Airport Inn is still there, reportedly serving the best pizza on the island. (Author's collection.)

On February 1, 1960, Lt. Donald Rumsfeld, USNR, came to Grosse Ile to drill on weekends. He served as a S2F aircraft commander with antisubmarine squadron VS-662, staying until October 31, 1960, when he was reassigned to Glenview Naval Air Station. Grosse Ile's squadrons were kept busy during the late 1950s and early 1960s. In the summer of 1958, a Navy reserve patrol squadron was on a two-week active-duty deployment in the Mediterranean when open revolt broke out against the pro-Western government of Lebanon. The base's five Neptunes, two R4Ds, and 121 reservists helped patrol the area until returning on July 28. In October 1961, Antisubmarine Squadron VS-733 was called to active duty during the Berlin Wall crisis. Reporting to NAS South Weymouth in Massachusetts, they did not return until July 1962.

In the 1960s, the base AVTECH Building had a mascot, an energetic little fellow who had the run of the place (and the nearby grounds). The pup was hanging around the building when he was adopted by two servicemen. This is the only known photograph of him. In recent years, the image hung outside the airport office with the caption, "Does anyone know his name?" As the years passed, it seemed more and more likely that the little fellow's name was lost to history. Then, one day in 2007, a veteran of the base walked into the airport office. "I remember that little fellow," he said. "What's his name?" everyone asked. "Dogface," the veteran said—and so the mystery was solved.

MICHIGAN
ARMY NATIONAL GUARD
NIKE SITE

U.S. ARMY

YOUR COMMUNITY'S ARMY AIR DEFENSE

The Army came to NASGI in 1955. In January, the Army claimed a small area of land north of the lower base as the site of a Nike Ajax missile instillation. Three silos were constructed, along with an underground storage bunker and an assembly and fueling area. The missile's tracking and control center was built on an acre of land on the north side of Groh Road. The Nike site became operational in January 1955 and was manned by Army crews who lived and ate in the Navy barracks. Designated Site D-51, it was one of several Nike batteries encircling Detroit. Between 22 and 30 Nike Ajax missiles were kept at the base, ready to fire on incoming Soviet bombers; each launcher could be reloaded in about a minute. The site closed in February 1963 and was never upgraded to the nuclear-armed Hercules missile (no nuclear weapons were ever based at NASGI). Today, the launch site is a nature preserve operated by the US Fish and Wildlife Service. The tracking and control area is now a landscaping business. The site's two concrete radar towers finally came down in 1998. This sign was found near Groh Road in the 1990s.

In 1959, the GCA unit at Grosse Ile was looking forward to its 40,000th ground-controlled approach since being activated over a dozen years earlier. Elaborate ceremonies were planned to mark the record-setting landing. Finally, the milestone aircraft approach began, but it was a transport plane from the Michigan Air National Guard. Should it have been given a go-around so the honor could have been bestowed upon a Navy crew? Nope, that is not the Navy way. The ANG C-47 was guided straight in, and its bewildered crew was appropriately honored.

In August 1950, the lower base was turned over to a search-and-rescue unit. Throughout the 1950s, a 140-foot rescue boat and its five-man crew were based in the former seaplane hangar. The approach was kept dredged 25 feet wide to allow the boat easy access to Lake Erie. It was also home to many pleasure boats belonging to reservists.

At the close of 1959, members of the search-and-rescue unit assembled a pontoon boat from scrap parts. Christened *Keller's Kanoe*, the homebuilt craft was named in honor of base commanding officer Capt. C.A. Keller. It served as a party boat for base personnel and was followed by *Reynold's Reck* in 1960 and *Schultz's Sloop* in 1962.

Tragedy struck in the early hours of Monday, December 9, 1968, when fire engulfed the former seaplane hangar. The blaze was spotted on the base shortly after it started, and Navy fire crews raced to the scene. Inside were over 40 boats belonging to base servicemen, as well as one of the base's party boats. By the time the fire crews arrived, the hangar was fully engulfed and beyond saving. The flames could be seen from as far away as Amherstburg and Luna Pier. An investigation listed the cause of the fire as suspicious, though no one was ever prosecuted.

The Navy left Grosse Ile in excellent condition. The Department of Defense declared that NASGI be closed by September 1967. The Skyraiders were gone, having quietly left for Southeast Asia in July 1965. All that remained were 16 squadrons of S2F Stoofs, SH34J Seabat antisubmarine helicopters, and a few Marine R4Q Flying Boxcars. Circumstances pushed the closing back to July 1969. In early December 1968, a squadron of Marine OV-10 Broncos arrived. Built in nearby Columbus, Ohio, the Bronco was the last new aircraft assigned to the base. The target date for the move came and went. The base celebrated its 40th birthday. Then word arrived, saying it would close in November. The last classes were held on November 2, 1969. The base's squadrons were reassigned to their new home, Selfridge Air National Guard Base, which was the same airfield where Torpedo Squadron VT-31 had first flown in 1927. The last aircraft to depart NASGI was a lone R5D that had been down with engine trouble. It finally departed in late November, bound for an overhaul center in Dothan, Alabama. As 1969 came to a close, all that remained on base was a small Navy security force. Despite the base closing, NASGI distinguished itself to the end, winning the Chief of Naval Air Trophy in 1969. But it was not the end for "Detroit's unsinkable aircraft carrier."

Five

ACCIDENTS

Accidents are an inevitable part of any training program. When they involve military aircraft, the results are usually dramatic. Over the years, Naval Air Station Grosse Ile had its share of accidents. Most occurred during training. This chapter is a collection of photographs of crashes that occurred at and around NASGI. They are interesting for the variety of aircraft involved, with some being rather unusual. They also show a wide range of outcomes—some were comic, but most tragic. They are not presented to horrify or glamorize the events; rather they depict another side of base life that, though rarely talked about, was very real.

Three additional accidents are also worth noting, though not depicted for lack of photographs. The only jet ever to crash on the base was not a Navy plane. On March 15, 1961, a Michigan Air National Guard RF-84F reconnaissance jet piloted by Lt. George Chekmakian crashed while attempting an emergency landing. The Thunderflash had lost its engine over Lake Erie while returning from Puerto Rico and was attempting an unpowered landing from the east on Runway 9-27. It crashed on the eastern end of the runway. Both pilot and plane were lost.

In the late 1940s, a pilot bailed out of his stricken FG-1D Corsair over East River Road. His parachute snagged in the trees in front of St. James Episcopal Church, leaving him dangling in front of the church. Today, his parachute release handle is on display in St. James. His Corsair is somewhere in Lake Erie.

Finally, on July 16, 1958, a S2F Stoof crashed into Frenchman's Creek near Groh Road, claiming the lives of Lt. Donald Leroy Southworth, who was checking out Lt. Comdr. Alan Radford Dale. Both men were popular reserve pilots whose deaths are still mourned today by base veterans.

Crashing in a biplane was different from crashing in a metal aircraft, as seen in these two photographs. Much slower than most all-metal aircraft, they had a tendency to break apart on impact. At slow speeds pilots could often walk away from accidents that might have claimed their lives in a more powerful machine, making planes like the Stearman and N3N good choices for primary flight training. Here are photographs of two accidents where the crews walked away. The photograph above shows a Curtiss O2C-1 that lost its engine and made a hard landing in a farmer's field. The props are unbroken, indicating they were not turning at the time. Below, a Stearman has managed to find the base's firehouse. Both aircraft were repaired and returned to service.

A Grumman TBM (BuNo 24291) crashed near Flat Rock while attempting an emergency landing on September 2, 1945. NASGI maintained many small airfields scattered throughout the downriver area where pilots could land in an emergency. Capt. S.T. Sandecki was trying to reach Nan-Bar Airport in Flat Rock when he went down. Nan-Bar was named for its owner's two daughters, Nancy and Barbara. The aircraft was scrapped.

A Curtiss SB2C-1 Helldiver (BuNo 18496) belonging to VB-97 crashed near the runway on September 5, 1945. The aircraft had skidded on ice while landing and was unable to brake in time. It finally came to rest after crashing through the airfield's outer fencing. The aircraft was repaired.

Douglas SBD-5 Dauntless dive-bomber No. 35 is raised by a crane after part of its landing gear collapsed while landing on Runway 35 on March 7, 1945. This aircraft (BuNo 36365) was a late model Dauntless belonging to a squadron returning from the Pacific. Obsolete by the time this photograph was taken, Dauntless aircraft were being phased out of service.

A four-bladed SB2C-3 Helldiver that ran off the runway and came to rest in Olds Bay is surveyed by a base firefighter standing on the plane's port wing. This aircraft (BuNo 20313) crashed on July 6, 1945, just days after returning to service from a previous crash. This time the aircraft was scrapped.

These photographs show the tragic loss of Vought F4U-4 Corsair (BuNo 56199) on June 18, 1945. Lt. D.W. Peebles was attempting to land when he stalled, which caused his port wing to strike the ground and the aircraft to cartwheel. During the crash, the Corsair's R-2800 engine was thrown free, coming to rest over a dozen yards away. Note the checkerboard blimp hangar in the background. The blimp hangar went through many paint schemes but wore the bright red checkerboard for most of its existence. The pilot of this aircraft was seriously injured. The aircraft was stricken.

This Vought F4U Corsair (BuNo 50559) was assigned to VBF-97, an East Coast Corsair training squadron assigned to NASGI. It crashed on July 10, 1945. The aircraft was piloted by Ens. F. C. Boyce when it suffered an engine failure. The pilot landed it safely in a field about 10 miles southwest of Grosse Ile. The aircraft was repaired.

This Navy F6F-3 Hellcat crashed into a farmer's field on Grosse Ile on July 17, 1945. The engine had quit, and the pilot was attempting to make an emergency landing at the base when the plane went down. The Hellcat's pilot escaped with only minor injuries before a fire broke out that consumed the plane. The aircraft was stricken.

This NFG-1D Corsair (BuNo 88124) crashed into a farmer's field near Gibralter after experiencing an engine failure on March 29, 1947. The FG-1D was a version of the Corsair built by the Goodyear Tire and Rubber Company at their plant in Akron, Ohio. By 1947, Corsairs on Grosse Ile no longer belonged to squadrons just passing through; all were assigned to the station. The plane (BuNo 88124) had an interesting history. It formerly belonged to fighter squadron VMF-512, which flew it from the escort carrier USS *Gilbert Islands* (CVE-107). While on the *Gilbert Islands*, the plane saw time in combat attacking Japanese targets in the Sakishima Gunto. Following this crash, the aircraft was deemed irreparable and scrapped.

A pair of crashed Martin AM-1 Maulers belonging to Naval Air Station Grosse Ile are seen in these two photographs; the one above taken on June 26, 1952 and below on July 15, 1953. Mauler 155 (above) crashed at the base when part of its landing gear collapsed during landing. It was not seriously damaged and was quickly returned to service. Mauler 138 (below) crashed into a field north of the base after experiencing engine problems. It also was not seriously damaged. Most Mauler accidents were caused by engine problems. The aircraft's 28-cylinder Pratt & Whitney R-4360 Double Wasp engines were complicated and prone to overheating. They were the main reason Maulers were quickly phased out of frontline service in favor of the smaller, more reliable AD-1 Skyraider.

Two F4U Corsairs meet the wrong way. The aircraft on the left, Corsair 12, was taxiing on the ramp in front of Hangar Two when it struck parked Corsair 8. This accident occurred in late 1948. Corsair 12's turning props have nearly severed the engine from Corsair 8, as well as slicing a huge gash in the aircraft's port wing.

In late 1953, the Douglas AD-1 Skyraider replaced Grosse Ile's Maulers, just as they had done in active Navy squadrons. Here a Skyraider (BuNo 26933) has come down in a farmer's field south of Gibralter. Reservists at NASGI could make emergency landings at several small downriver airports, though sometimes the planes just could not make it.

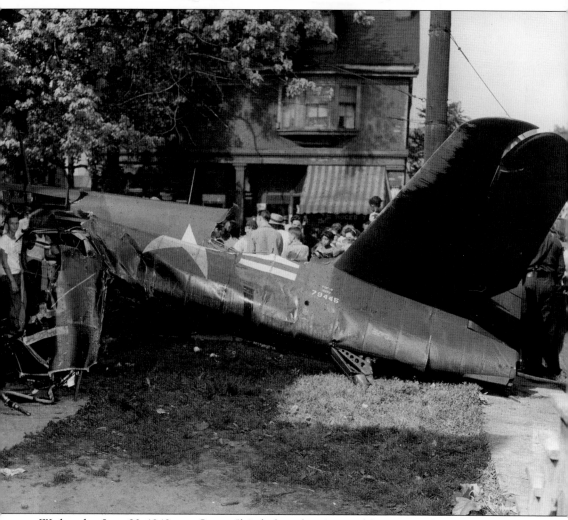

Wednesday, June 23, 1948, was Grosse Ile's darkest day. A mixed formation of seven navy reserve F4U-4 Corsairs and F6F Hellcats were flying over downtown Detroit. The flight leader later stated that he observed one of the planes "slide across in front of the other" in the formation while making a right-hand turn, causing the two aircraft to collide in mid-air. During the collision, a F4U's prop severed the tail from a F6F. The F6F-5 Hellcat (BuNo 79445) was piloted by Ens. H.J. Nicholson, while the F4U-4 Corsair (BuNo 96790) was piloted by Ens. R.K. Schultheiss. The two men killed were participating in a training program for navy reserve pilots from attack squadron 89-A, based on Grosse Ile. The pilots could have saved themselves and bailed out from their planes, parachuting to safety, but they chose to guide the aircraft away from a heavily populated area along Gratiot Avenue. Ensign Nicholson's Hellcat, seen on this page, plunged through the roof of the Putnam Tool Company on Charlevoix Street. All 60 employees escaped unharmed. Ensign Schultheiss's Corsair crashed and exploded two blocks away on Benson Street. The wreckage set four homes on fire and caused several civilian injuries.

Six

GROSSE ILE TODAY

When a naval air station closes, a series of events usually occur to mark its passing. It starts when the last plane flies out. The station keepers then say their last good-byes before moving on to their next assignments. Finally, within a few months, the bulldozers arrive. Soon, all that remains of the once-proud air station are photographs, records, and the memories of the men and women who were stationed there. But a funny thing happened when Naval Air Station Grosse Ile closed in 1969—the bulldozers never arrived. Instead, the Grosse Ile Township Board of Trustees voted to turn the base into the Grosse Ile Township Airport. Hangars that had recently contained R4Q Flying Boxcars and OV-10 Broncos were soon filling with Bonanzas, Cessnas, and Republic Seabees. The administrative wing of Hangar One became staffed with officials who did not wear Navy uniforms. The old Officers' Club became a busy restaurant accurately named the Officers' Club. And Groh Road, long off-limits to island residents, was reopened, finally reconnecting the north and south sides of Grosse Ile. Over the years, much has changed. All the buildings that sprung up in 1941 north of Groh Road have come down, including the much-beloved recreation hall. The lower base is a nature preserve, slowly returning to a cattail marsh. The mammoth drill hall is filled with tennis courts, and the EPA has turned the historic Tin Hangar into a storage shed. Only the Airport Inn at the intersection of Groh and Meridian Roads has remained largely unchanged, with its windows now filled with stained-glass renderings of vintage war birds. At Grosse Ile, a student pilot can still learn to fly from the very same runways where young British and American airmen once took to the skies in Stearman biplanes before going off to fight in World War II. We are all fortunate that the Grosse Ile Township Airport will remain to continue the legacy of the thousands of men and women who served there, helping to keep the world safe for our generation and the ones to follow.

This photograph shows Hangar One (foreground) and Hangar Two (background) in early May 2011. The closing of NASGI caused a frenzy of activity from groups interested in obtaining some or the entire base. The Coast Guard wanted three to five acres for a station, while the Wayne County Road Commission wanted the base for a county airport. Ultimately, it was the Grosse Ile Board of Trustees that had the final say on the base's disposal. In April 1967, the board hired a consulting firm to perform a study on the possibility of reopening the base as a township airport, and 19 months later they received a report showing that it was economically feasible. The plan was endorsed by the Wayne County Supervisor's Aircraft Committee, and in 1968 a vote approved a $750,000 bond plus federal and state aid. After minor improvements, the base reopened in a ceremony held on July 2, 1972. From then on it would be known as the Grosse Ile Township Airport. (Author's collection.)

The Grosse Ile Board of Trustees was not just purchasing an airport, it was receiving an entire community of buildings. The facilities in Hangar One's administrative section were an improvement over the township's old offices, and in 1999 a decision was made that the township's entire staff be relocated to offices on the former base. Hangar One's west administrative offices, as well as the administrative wing, were completely gutted and remodeled. In the fall of 2000, the entire staff and administration of Grosse Ile moved to the former base. Today, the township staff, as well as the airport's administrative staff, occupies offices on the hangar's west side. The administration wing has been turned into the Naval Air Station Grosse Ile Lobby Museum, and the second floor was recently renovated as the township's boardroom. Hangar One is now essentially the Grosse Ile Township Hall. (Author's collection.)

The township's recently renovated boardroom occupies virtually the entire upper floor of Hangar One's administration wing. The boardroom has the latest in interactive displays, as well as flat-panel televisions for the benefit of audience members. The township's seal is displayed behind the board's seats. On the walls around the room are oil paintings depicting vintage aircraft. They were painted by John Ansteth, a pilot whose grandfather was the original pilot of the ZMC-2. (Both author's collection.)

During the renovation in 1999, space was set aside in the old administrative wing for the Grosse Ile Historical Society (GIHS) to create a museum (above), which opened in 2001. A call went out to base veterans for any artifacts, uniforms, articles, or photographs they were willing to donate to the planned museum. They received such an abundance of material that much of it could not be included in the initial displays. In 2002, a Memorial Garden was installed behind the museum by the Grosse Ile Kiwanis (below). It has been the site of memorial ceremonies for several veterans. Today, the museum is operated by the GIHS, while the Memorial Garden continues to be maintained by the Grosse Ile Kiwanis. The GIHS continues to receive artifacts from the former base on a regular basis. In 2011, an anonymous donor sent the historical society one of the control tower's original World War II signal flare guns.

The interior of Hangar One, as it appeared in May 2011, is not very different from its appearance in 1942. Most of the windows and roof panels are original. Amazingly, the occupants are mostly World War II vintage as well. This photograph and the two on the next page show the flying collection of the Yankee Air Museum, which calls Hangar One home. Tragedy struck on the night of October 9, 2004, when a fire broke out that consumed the museum's huge hangar at Willow Run Airport, destroying its collection of artifacts and several of its vintage aircraft. Museum staff members risked their lives pushing the museum's four flyable aircraft out with their bare hands before the building collapsed. (Author's collection.)

Today, those four aircraft saved from the 2004 fire, which include the Boeing B-17G s/n 44-85829 *Yankee Lady*, North American B-25D s/n 43-3634 *Yankee Warrior*, Douglas C-47 *Yankee Doodle Dandy*, and Stinson V-77 Reliant *Yankee Friendship*, are stored in Hangar One when not appearing at air shows. They will likely remain at Grosse Ile until the museum's new home is completed at Willow Run. In 2010, Hangar One was also used as a set for the futuristic Hugh Jackman film *Real Steel*. (Both author's collection.)

Hangar Two, as it appears today, is also little changed from its earliest days. Built in 1927, it was the second building constructed on-site after the blimp hangar. Over the years, it has been home to two flying schools before becoming the Navy's main hangar until replaced by Hangar One. Today it is still an aircraft hangar and is used by area residents to store their planes—though the aircraft found inside are decidedly more modern than the Grumman FF-2s and Corsairs that once parked here. Today, the oldest resident is an aging Aero Commander. Other current residents include a Russian Yak-52 trainer (above). It is ironic that Hangar Two, once home to planes designed to defeat the Russians, is today home to Russian aircraft. (Both author's collection.)

The drill hall is one of the best maintained and busiest of the base's surviving buildings. In the 1970s, it was turned into the Grosse Ile Tennis Courts, a huge indoor tennis facility where island residents can practice their game during the worst Michigan winters. These days, the only inspections held inside concern a player's tennis swing. (Author's collection.)

The Officers' Club, seen here in May 2011, has found a continuing way of life not so different from the one for which it was built. Constructed in 1927 as a barracks for the flying school, and later used as the Navy's barracks until World War II, the building has seen the changing fortunes of more businesses than any other structure on base. In recent years, it has become home to the Pilot House, a fashionable hotel and restaurant. (Author's collection.)

Today, the Grosse Ile Township Airport is one of the busiest airports in southeastern Michigan. It is home to nearly 100 private aircraft and a museum collection of World War II war birds. It averages about 24,000 air operations a year. Veterans and historians can count on a future that can only become brighter. Naval Air Station Grosse Ile started out as an experiment in metal-clad airships. It was transformed by the winds of war into one of the country's most important training bases. It sent its pilots to fight in the skies over Korea. It fought the Cold War with diligence and has finally surrendered itself to a comfortable life in retirement—no one can ask more of it than that. Like the veterans who lived and trained there, it gave its best, and it will continue to do so in whatever role is asked of it. The sun has not yet set, and the last stories have not yet been written for "Detroit's unsinkable aircraft carrier." (Author's collection.)

www.arcadiapublishing.com

MAP SEARCH

Discover books about the town where you grew up, the cities where your friends and families live, the town where your parents met, or even that retirement spot you've been dreaming about. Our Web site provides history lovers with exclusive deals, advanced notification about new titles, e-mail alerts of author events, and much more.

MADE IN THE USA

Arcadia Publishing, the leading local history publisher in the United States, is committed to making history accessible and meaningful through publishing books that celebrate and preserve the heritage of America's people and places. Consistent with our mission to preserve history on a local level, this book was printed in South Carolina on American-made paper and manufactured entirely in the United States.

This book carries the accredited Forest Stewardship Council (FSC) label and is printed on 100 percent FSC-certified paper. Products carrying the FSC label are independently certified to assure consumers that they come from forests that are managed to meet the social, economic, and ecological needs of present and future generations.

FSC

Mixed Sources
Product group from well-managed forests and other controlled sources

Cert no. SW-COC-001530
www.fsc.org
© 1996 Forest Stewardship Council

Find Your Place in History.